Editor
Gisela Lee

Managing Editor
Karen Goldfluss, M.S. Ed.

Editor-in-Chief
Sharon Coan, M.S. Ed.

Cover Artist
Barb Lorseyedi

Art Coordinator
Kevin Barnes

Art Director
CJae Froshay

Imaging
Alfred Lau
James Edward Grace

Product Manager
Phil Garcia

Publishers
Rachelle Cracchiolo, M.S. Ed.
Mary Dupuy Smith, M.S. Ed.

Author

Teacher Created Materials Staff

Teacher Created Materials, Inc.
6421 Industry Way
Westminster, CA 92683
www.teachercreated.com
ISBN-0-7439-3316-8

Table of Contents

Introduction

The old adage "practice makes perfect" can really hold true for your child and his or her education. The more practice and exposure your child has with concepts being taught in school, the more success he or she is likely to find. For many parents, knowing how to help their children can be frustrating because the resources may not be readily available. As a parent it is also difficult to know where to focus your efforts so that the extra practice your child receives at home supports what he or she is learning in school.

This book has been designed to help parents and teachers reinforce basic skills with their children. *Practice Makes Perfect* reviews basic math skills for children in the second grade. The math focus is on addition and subtraction. While it would be impossible to include all concepts taught in the second grade in this book, the following basic objectives are reinforced through practice exercises. These objectives support math standards established on a district, state, or national level. (Refer to the Table of Contents for the specific objectives of each practice page.)

- addition and subtraction facts to 10
- word problems
- two-digit addition and subtraction with regrouping
- two-digit addition and subtraction without regrouping

There are 36 practice pages organized sequentially, so children can build their knowledge from more basic skills to higher-level math skills. To correct the practice pages in this book, use the answer key provided on pages 47 and 48. Six practice tests follow the practice pages. These provide children with multiple-choice test items to help prepare them for standardized tests administered in schools. As children complete a problem, they fill in the correct letter among the answer choices. An optional "bubble-in" answer sheet has also been provided on page 46. This answer sheet is similar to those found on standardized tests. As your child completes each test, he or she can fill in the correct bubbles on the answer sheet.

How to Make the Most of This Book

Here are some useful ideas for optimizing the practice pages in this book:

- Set aside a specific place in your home to work on the practice pages. Keep it neat and tidy with materials on hand.
- Set up a certain time of day to work on the practice pages. This will establish consistency. An alternative is to look for times in your day or week that are less hectic and conducive to practicing skills.
- Keep all practice sessions with your child positive and constructive. If the mood becomes tense, or you and your child are frustrated, set the book aside and look for another time to practice with your child. Forcing your child to perform will not help. Do not use this book as a punishment.
- Help with instructions if necessary. If your child is having difficulty understanding what to do or how to get started, work the first problem through with him or her.
- Review the work your child has done. This serves as reinforcement and provides further practice.
- Allow your child to use whatever writing instruments he or she prefers. For example, colored pencils can add variety and pleasure to drill work.
- Pay attention to the areas in which your child has the most difficulty. Provide extra guidance and exercises in those areas. Allowing children to use drawings and manipulatives, such as coins, tiles, game markers, or flash cards, can help them grasp difficult concepts more easily.
- Look for ways to make real-life application to the skills being reinforced.

Practice 1

Solve each problem.

1.
3
4
3
+ 2

2.
4
3
2
+ 1

3.
2
5
2
+ 2

4.
4
1
2
+ 1

5.
1
3
1
+ 1

6.
2
1
3
+ 1

7.
3
2
1
+ 3

8.
2
4
+ 2

9.
7
2
+ 1

10.
7
4
+ 1

11.
7
2
3
1
+ 1

12.
3
5
+ 1

13.
6
2
+ 3

14.
9
2
1
+ 1

15.
3
4
3
+ 2

16.
8
3
2
3
+ 1

17.
9
1
4
+ 2

18.
8
2
4
5
+ 1

19.
6
3
2
+ 4

20.
5
2
7
9
+ 1

Practice 2

Solve each problem.

1.

10 + 10 = ______
10 + 11 = ______
10 + 12 = ______
10 + 13 = ______
10 + 14 = ______
10 + 15 = ______
10 + 16 = ______
10 + 17 = ______
10 + 18 = ______
10 + 19 = ______
10 + 20 = ______

2.

20 + 10 = ______
20 + 11 = ______
20 + 12 = ______
20 + 13 = ______
20 + 14 = ______
20 + 15 = ______
20 + 16 = ______
20 + 17 = ______
20 + 18 = ______
20 + 19 = ______
20 + 20 = ______

3.

30 + 10 = ______
30 + 11 = ______
30 + 12 = ______
30 + 13 = ______
30 + 14 = ______
30 + 15 = ______
30 + 16 = ______
30 + 17 = ______
30 + 18 = ______
30 + 19 = ______
30 + 20 = ______

4.

40 + 10 = ______
40 + 11 = ______
40 + 12 = ______
40 + 13 = ______
40 + 14 = ______
40 + 15 = ______
40 + 16 = ______
40 + 17 = ______
40 + 18 = ______
40 + 19 = ______
40 + 20 = ______

5.

50 + 10 = ______
50 + 11 = ______
50 + 12 = ______
50 + 13 = ______
50 + 14 = ______
50 + 15 = ______
50 + 16 = ______
50 + 17 = ______
50 + 18 = ______
50 + 19 = ______
50 + 20 = ______

6.

60 + 10 = ______
60 + 11 = ______
60 + 12 = ______
60 + 13 = ______
60 + 14 = ______
60 + 15 = ______
60 + 16 = ______
60 + 17 = ______
60 + 18 = ______
60 + 19 = ______
60 + 20 = ______

7.

70 + 10 = ______
70 + 11 = ______
70 + 12 = ______
70 + 13 = ______
70 + 14 = ______
70 + 15 = ______
70 + 16 = ______
70 + 17 = ______
70 + 18 = ______
70 + 19 = ______
70 + 20 = ______

8.

80 + 10 = ______
80 + 11 = ______
80 + 12 = ______
80 + 13 = ______
80 + 14 = ______
80 + 15 = ______
80 + 16 = ______
80 + 17 = ______
80 + 18 = ______
80 + 19 = ______
80 + 20 = ______

Practice 3

Find the sums. Then follow the directions and color the picture.

Use a red crayon to color blocks with sums of 24.

Use a yellow crayon to color blocks with sums of 38.

Use a blue crayon to color blocks with sums of 41.

Use a green crayon to color blocks with sums of 57.

1. $\begin{array}{r} 20 \\ +\ 21 \\ \hline \end{array}$	**2.** $\begin{array}{r} 26 \\ +\ 12 \\ \hline \end{array}$	**3.** $\begin{array}{r} 12 \\ +\ 12 \\ \hline \end{array}$
4. $\begin{array}{r} 32 \\ +\ 25 \\ \hline \end{array}$	**5.** $\begin{array}{r} 14 \\ +\ 10 \\ \hline \end{array}$	**6.** $\begin{array}{r} 15 \\ +\ 42 \\ \hline \end{array}$
7. $\begin{array}{r} 11 \\ +\ 13 \\ \hline \end{array}$	**8.** $\begin{array}{r} 15 \\ +\ 23 \\ \hline \end{array}$	**9.** $\begin{array}{r} 11 \\ +\ 30 \\ \hline \end{array}$

Practice 4

Find the sum of the numbers in each circle.

1. $21 + 12$

2. $16 + 13$

3. $32 + 11$

4. $10 + 19$

5. $21 + 13$

6. $31 + 22$

7. $40 + 24$

8. $52 + 10$

9. $41 + 18$

Practice 5

To discover the secret number, find the sums and follow the directions.

1. 21 + 18	**2.** 31 + 16	**3.** 31 + 21
4. 41 + 31	**5.** 12 + 12	**6.** 10 + 17

☆ It is not number 24. Cross it out.

☆ It is not number 39. Cross it out.

☆ It is not number 52. Cross it out.

☆ It is not number 72. Cross it out.

☆ It is not number 47. Cross it out.

What is the secret number?______________

Practice 6           

Guess what is in the box. Find the sums. Then write the letter in each box that matches each sum. Read the word you spell and draw it in the box.

25	26	27	28	29
k	c	a	o	l

Example					
13 + 14 = 27	14 + 12	11 + 18	16 + 12	13 + 13	12 + 13
a		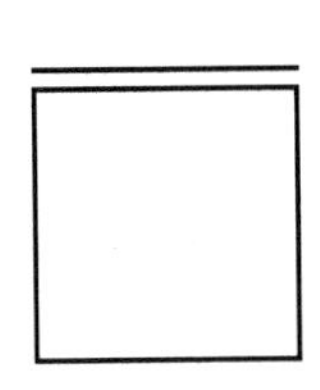	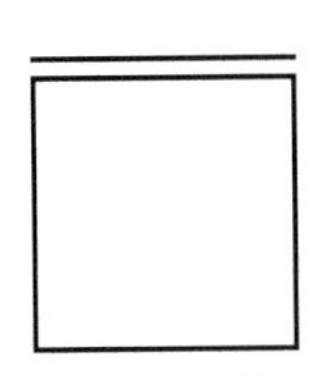	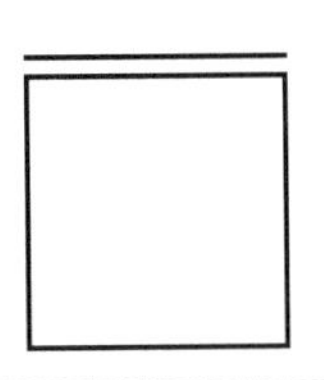	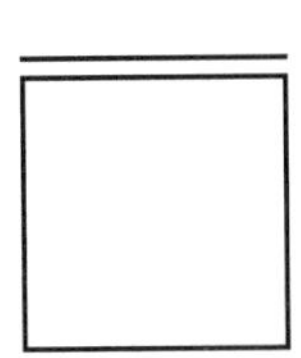

Practice 7

To discover a secret message, find the sums. Then place each letter in the matching numbered space.

A.	B.	C.	D.	E.	G.
21 + 21	23 + 12	51 + 10	15 + 14	10 + 10	26 + 11

I.	L.	M.	N.	O.	R.
31 + 20	42 + 11	51 + 41	53 + 31	24 + 43	21 + 10

S.	T.	U.	W.	Y.
33 + 33	13 + 12	24 + 15	34 + 14	62 + 10

___ ___ ___ ___ ___ ___ ___ ___ ___ ___ ___ ___ ___ ___ ___!
61 67 84 37 31 42 25 39 53 42 25 51 67 84 66

___ ___ ___   ___ ___ ___   ___ ___ ___ ___ ___ ___
72 67 39   42 31 20   42 29 29 51 84 37

___ ___ ___ — ___ ___ ___ ___ ___
25 48 67   29 51 37 51 25

___ ___ ___ ___ ___ ___ ___.
84 39 92 35 20 31 66

Practice 8

Guess what is in the box. Find the sums. Then write the letter in each box that matches each sum. Read the word you spell and draw it in the box.

48	49	50	51	52	53
s	e	a	o	h	r

23	26	14	39	27	30
+27	+26	+37	+14	+21	+19
50					
a	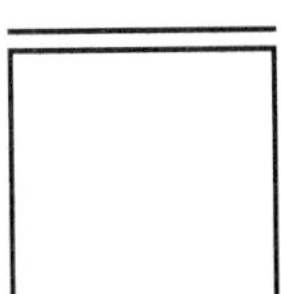	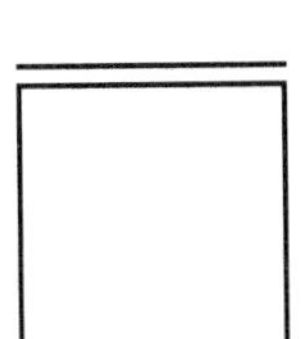	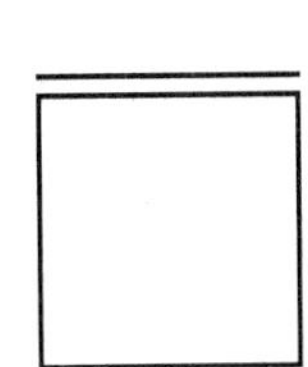	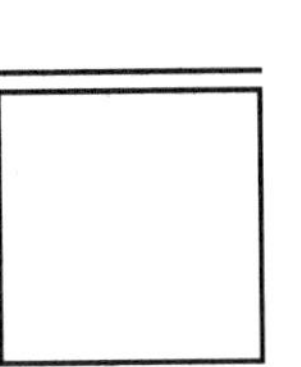	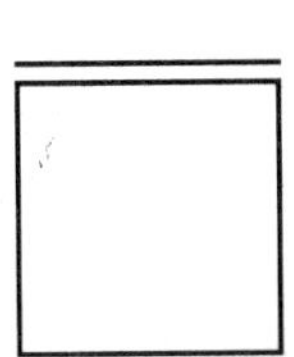

Practice 9

Find the sums of the numbers on the target. Where did Aaron's arrow land if he hit 60? Color that ring green.

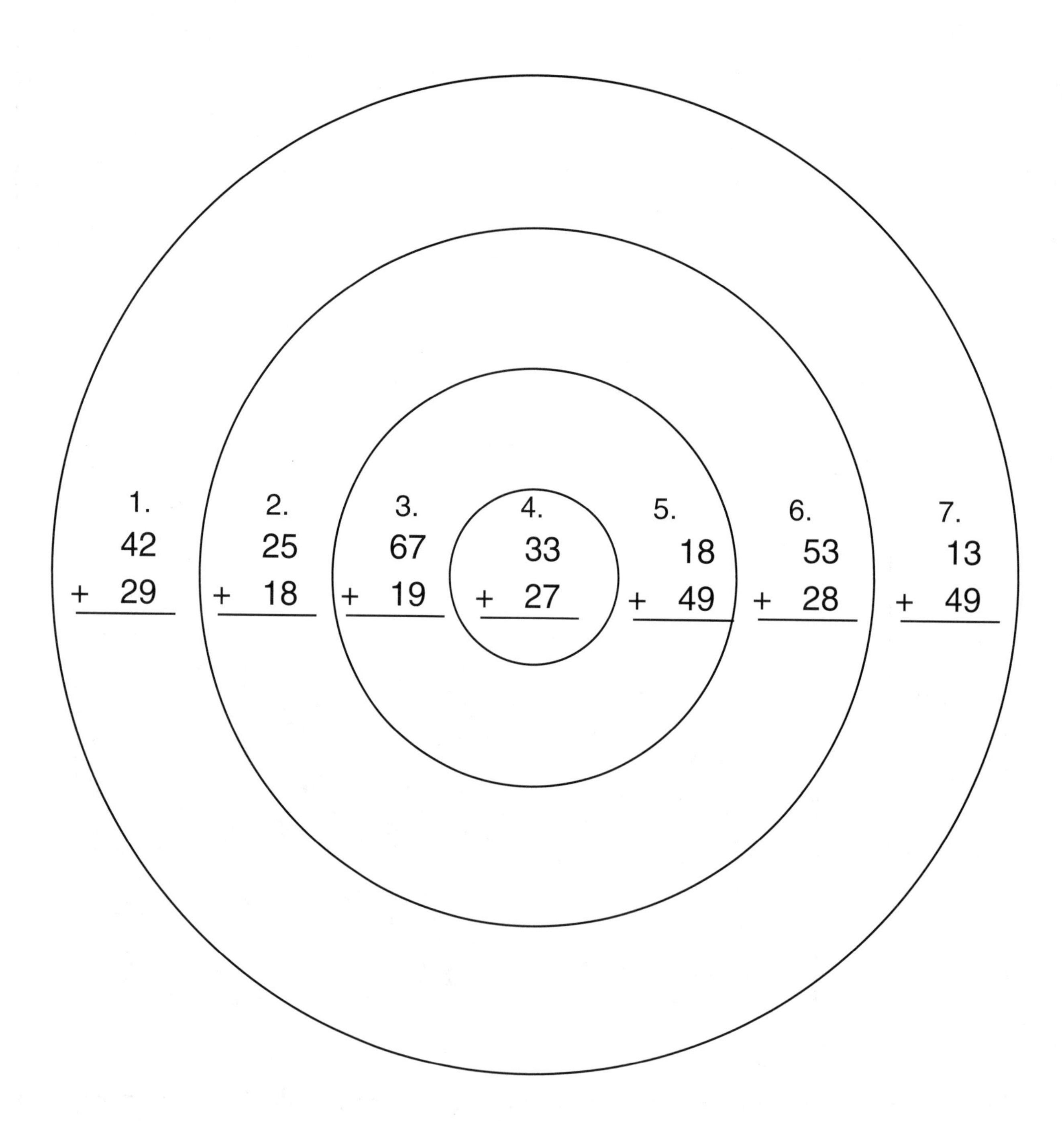

Practice 10

Guess what is in the box. Find the sums. Then write the letter in each box that matches each sum. Read the word you spell and draw it in the box.

80	81	82	83	84	85	86	87
d	e	a	o	c	r	l	i

$$\begin{array}{r} 45 \\ +37 \\ \hline 82 \end{array}$$

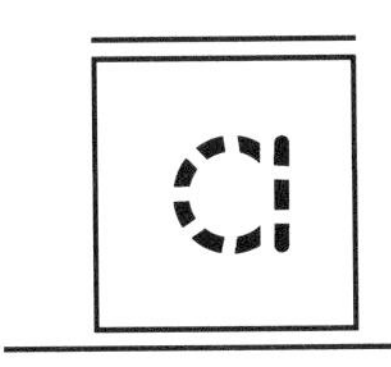

$$\begin{array}{r} 24 \\ +60 \\ \hline \end{array}$$

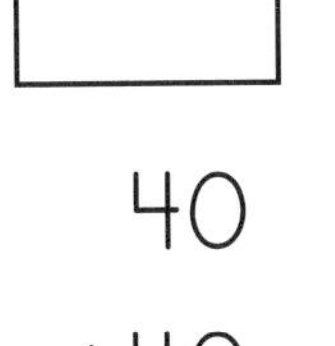

$$\begin{array}{r} 62 \\ +23 \\ \hline \end{array}$$

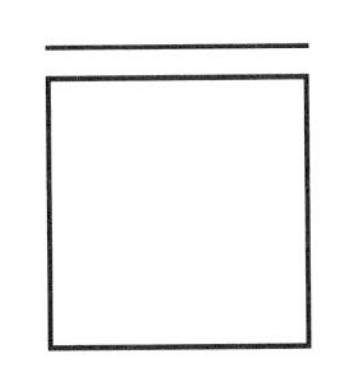

$$\begin{array}{r} 58 \\ +25 \\ \hline \end{array}$$

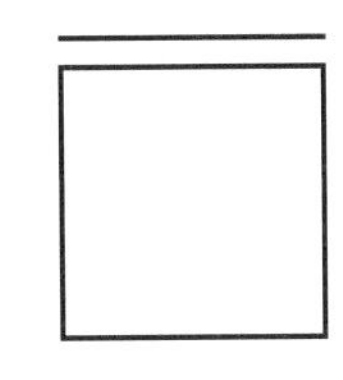

$$\begin{array}{r} 12 \\ +72 \\ \hline \end{array}$$

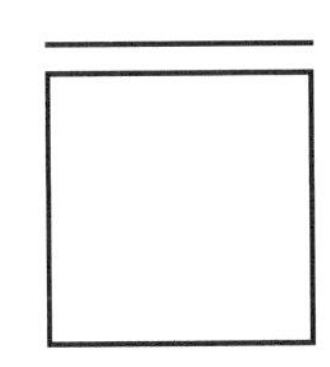

$$\begin{array}{r} 36 \\ +47 \\ \hline \end{array}$$

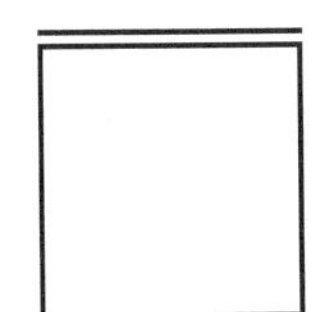

$$\begin{array}{r} 40 \\ +40 \\ \hline \end{array}$$

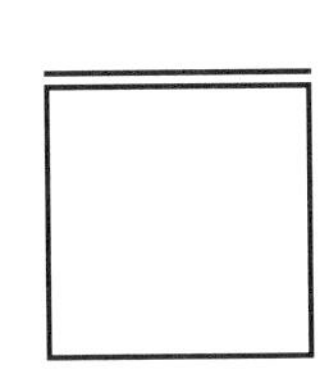

$$\begin{array}{r} 52 \\ +35 \\ \hline \end{array}$$

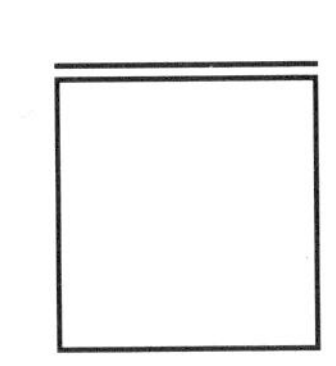

$$\begin{array}{r} 32 \\ +54 \\ \hline \end{array}$$

$$\begin{array}{r} 48 \\ +33 \\ \hline \end{array}$$

Practice 11

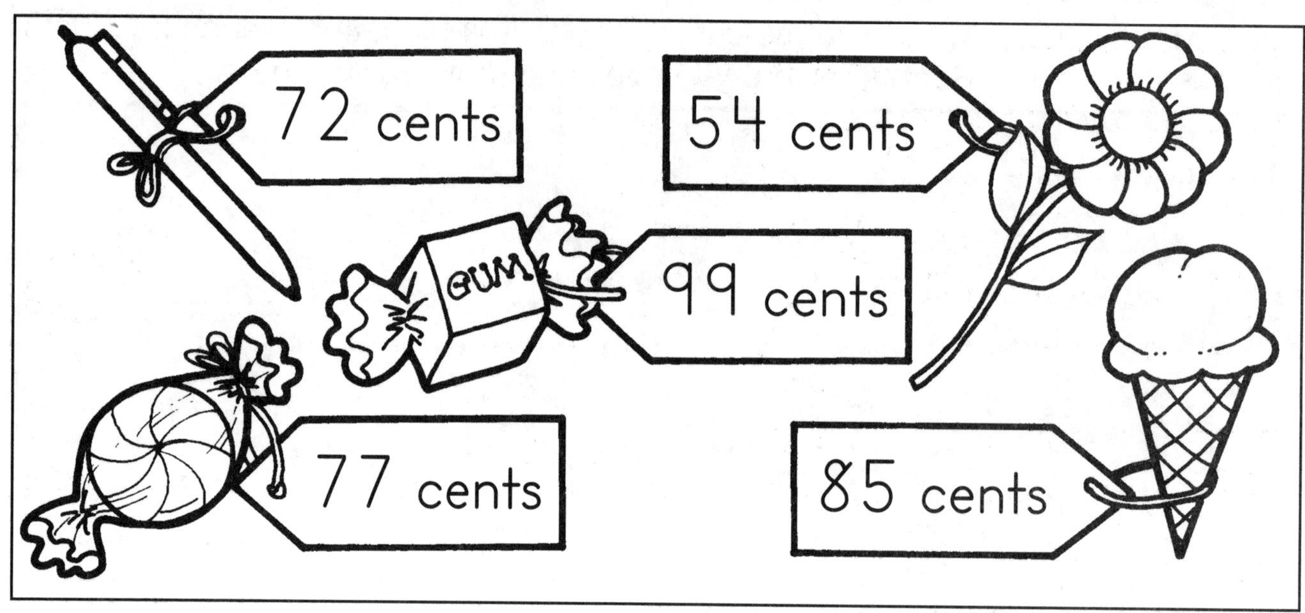

Use the prices to write addition problems. Find the sums.

1. | + | =

___ + ___ = ___ cents

4. | + | =

___ + ___ = ___

2. | + | =

___ + ___ = ___

5. | + | + | =

___ + ___ + ___ = ___

3. | + | =

___ + ___ = ___

6. | + | + | =

___ + ___ + ___ = ___

Practice 12

Use the prices to write addition problems. Find the sums.

1. (sweater) + (shoes) =

   ___ + ___ = ___ dollars

2. (dress) + (pants) =

   ___ + ___ = ___

3. (hat) + (watch) =

   ___ + ___ = ___

4. (pants) + (shoes) =

   ___ + ___ = ___

5. (watch) + (hat) + (sweater) =

   ___ + ___ + ___ = ___

6. (dress) + (shoes) + (pants) =

   ___ + ___ + ___ = ___

Practice 13

Read each word problem. In the box, write the number sentence it shows. Find the sum.

1

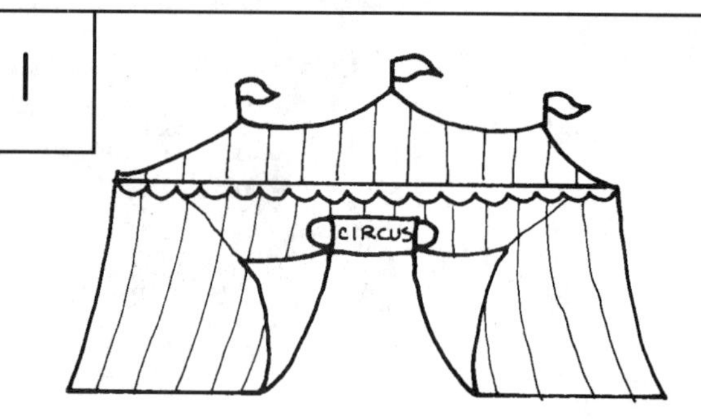

At the circus, Kenny saw 16 tigers, 14 horses, and 12 monkeys. How many animals did he see in all?

16 + 14 + 12 = 42

2

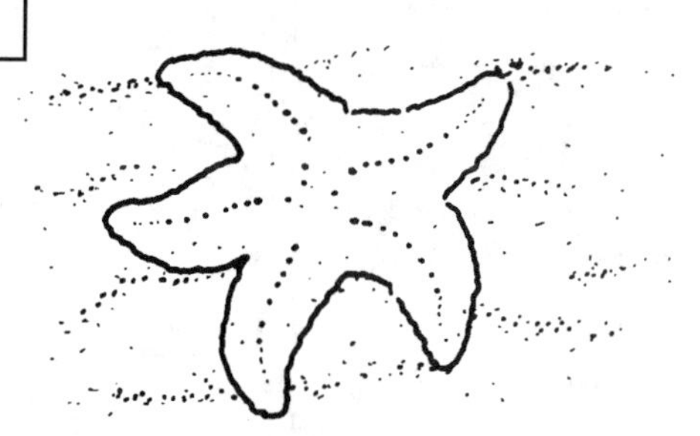

When Sandra went to the tidepools, she counted 18 starfish, 22 fish, and 36 shells. How many things did she see in all?

3

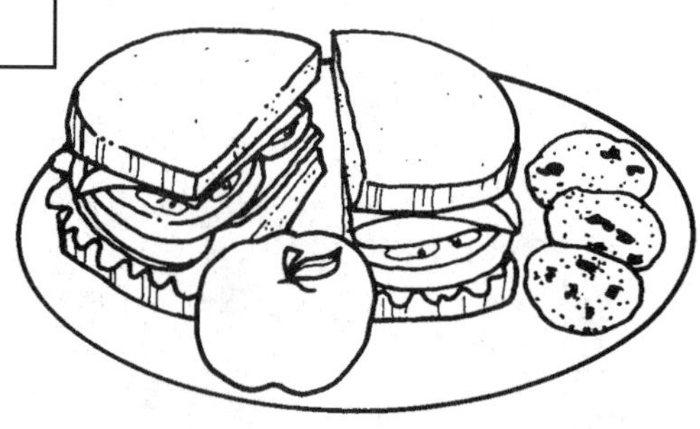

During one month, Jared ate 17 sandwiches, 13 apples, and 42 cookies. How many things did he eat in all?

4

Emily did 15 addition problems and 23 subtraction problems at school. At home, her mother gave her 24 more. How many problems did she solve in all?

Practice 14

Read each word problem. Write the number sentence it shows. Find the sum.

1

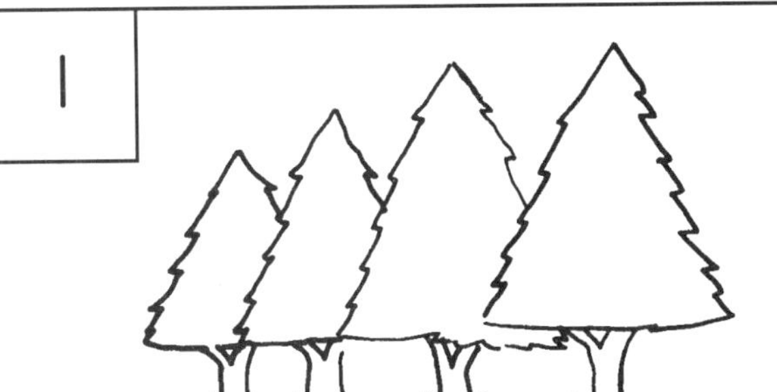

In the forest, Lisa counted 53 pine trees, 24 spider webs, and 12 chipmunks. How many things did she count in all?

53 + 24 + 16 = 89

2

In Bill's classroom there are 57 pencils, 24 pieces of chalk, and 43 bottles of glue. How many supplies are there in all?

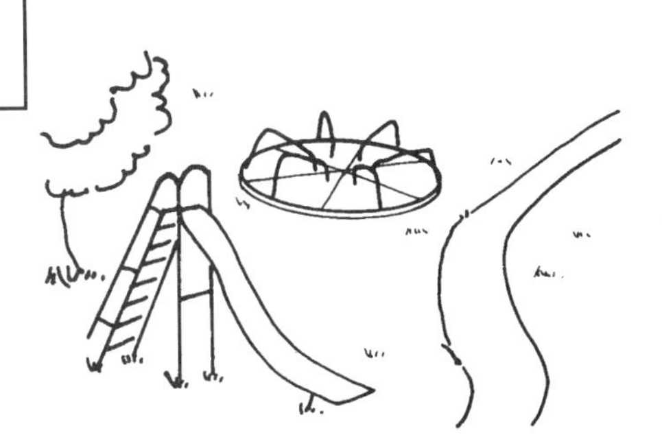

At the park, Carla counted 24 ducks, 32 children, and 34 roller skates. How many things did she count in all?

4

James counted 36 stars one night, 40 stars the next, and 67 on the third night. How many stars did he count in all?

Practice 15

Find the sums.

1. 29 + 48	7. 14 + 75	13. 74 + 68	19. 97 + 50
2. 37 + 95	8. 20 + 52	14. 38 + 15	20. 45 + 29
3. 10 + 36	9. 41 + 52	15. 25 + 49	21. 34 + 17
4. 56 + 26	10. 27 + 30	16. 39 + 27	22. 74 + 19
5. 40 + 33	11. 52 + 73	17. 10 + 64	23. 27 + 28
6. 86 + 56	12. 67 + 70	18. 86 + 16	24. 55 + 54

Practice 16

Find the sums.

1. 11 + 50	7. 69 + 12	13. 69 + 16	19. 36 + 13
2. 64 + 42	8. 72 + 38	14. 71 + 59	20. 29 + 80
3. 24 + 93	9. 48 + 18	15. 13 + 68	21. 51 + 17
4. 17 + 20	10. 52 + 11	16. 41 + 96	22. 19 + 91
5. 58 + 72	11. 15 + 19	17. 82 + 30	23. 31 + 46
6. 67 + 14	12. 31 + 62	18. 93 + 90	24. 87 + 43

Practice 17

Find the sums.

1. 21 + 33	7. 39 + 72	13. 25 + 81	19. 37 + 69
2. 75 + 42	8. 51 + 57	14. 51 + 20	20. 38 + 94
3. 24 + 53	9. 48 + 84	15. 12 + 48	21. 28 + 87
4. 42 + 26	10. 23 + 64	16. 51 + 76	22. 69 + 43
5. 68 + 82	11. 14 + 86	17. 32 + 36	23. 41 + 46
6. 61 + 33	12. 33 + 52	18. 97 + 60	24. 77 + 63

Practice 18

  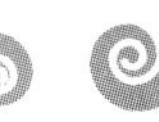        

Find the sums.

1. 39 57 + 47	7. 39 12 + 72	13. 26 71 + 59	19. 17 79 + 54
2. 33 75 + 23	8. 51 24 + 88	14. 52 30 + 18	20. 39 95 + 48
3. 21 53 + 17	9. 42 84 + 19	15. 13 38 + 42	21. 27 77 + 70
4. 42 26 + 49	10. 23 14 + 92	16. 52 38 + 42	22. 59 44 + 16
5. 68 62 + 56	11. 84 36 + 65	17. 52 66 + 83	23. 51 36 + 24
6. 61 33 + 63	12. 34 42 + 30	18. 98 61 + 15	24. 67 73 + 30

Practice 19

Solve each problem.

1.	2.	3.	4.
20 – 10 = ______	30 – 10 = ______	40 – 10 = ______	50 – 10 = ______
20 – 11 = ______	30 – 11 = ______	40 – 11 = ______	50 – 11 = ______
20 – 12 = ______	30 – 12 = ______	40 – 12 = ______	50 – 12 = ______
20 – 13 = ______	30 – 13 = ______	40 – 13 = ______	50 – 13 = ______
20 – 14 = ______	30 – 14 = ______	40 – 14 = ______	50 – 14 = ______
20 – 15 = ______	30 – 15 = ______	40 – 15 = ______	50 – 15 = ______
20 – 16 = ______	30 – 16 = ______	40 – 16 = ______	50 – 16 = ______
20 – 17 = ______	30 – 17 = ______	40 – 17 = ______	50 – 17 = ______
20 – 18 = ______	30 – 18 = ______	40 – 18 = ______	50 – 18 = ______
20 – 19 = ______	30 – 19 = ______	40 – 19 = ______	50 – 19 = ______
20 – 20 = ______	30 – 20 = ______	40 – 20 = ______	50 – 20 = ______

5.	6.	7.	8.
60 – 10 = ______	70 – 10 = ______	80 – 10 = ______	90 – 10 = ______
60 – 11 = ______	70 – 11 = ______	80 – 11 = ______	90 – 11 = ______
60 – 12 = ______	70 – 12 = ______	80 – 12 = ______	90 – 12 = ______
60 – 13 = ______	70 – 13 = ______	80 – 13 = ______	90 – 13 = ______
60 – 14 = ______	70 – 14 = ______	80 – 14 = ______	90 – 14 = ______
60 – 15 = ______	70 – 15 = ______	80 – 15 = ______	90 – 15 = ______
60 – 16 = ______	70 – 16 = ______	80 – 16 = ______	90 – 16 = ______
60 – 17 = ______	70 – 17 = ______	80 – 17 = ______	90 – 17 = ______
60 – 18 = ______	70 – 18 = ______	80 – 18 = ______	90 – 18 = ______
60 – 19 = ______	70 – 19 = ______	80 – 19 = ______	90 – 19 = ______
60 – 20 = ______	70 – 20 = ______	80 – 20 = ______	90 – 20 = ______

Practice 20          

Solve each subtraction problem. Then use the color code to color the picture.

1, 2 = blue
3, 4, 5, 6 = yellow
11, 12 = white
7, 8 = red
9, 10 = green

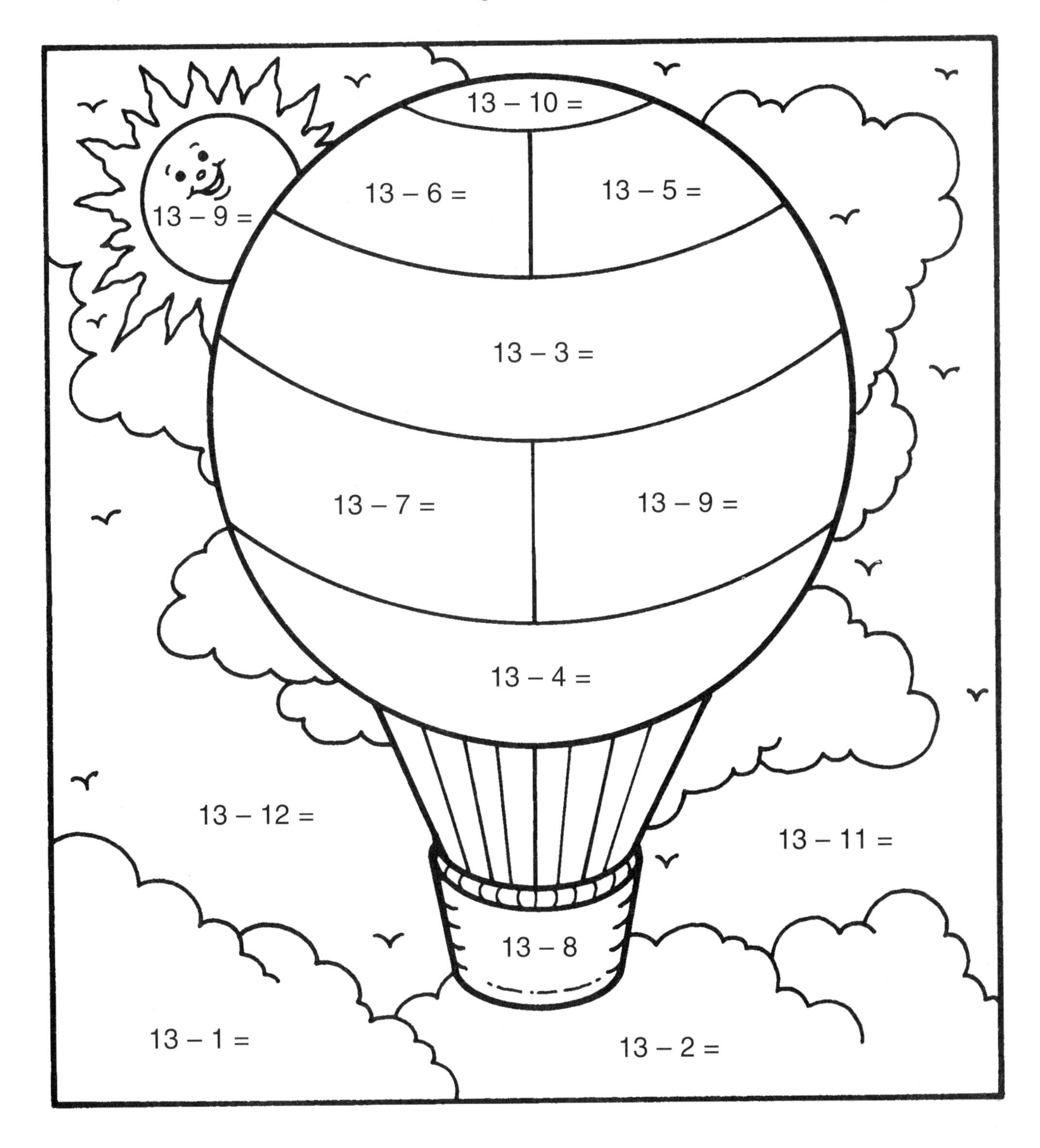

Practice 21

To discover what is hidden in the picture, use a blue crayon to color the areas with a difference more than 10. Use a gray crayon to color the areas with a difference less than 10.

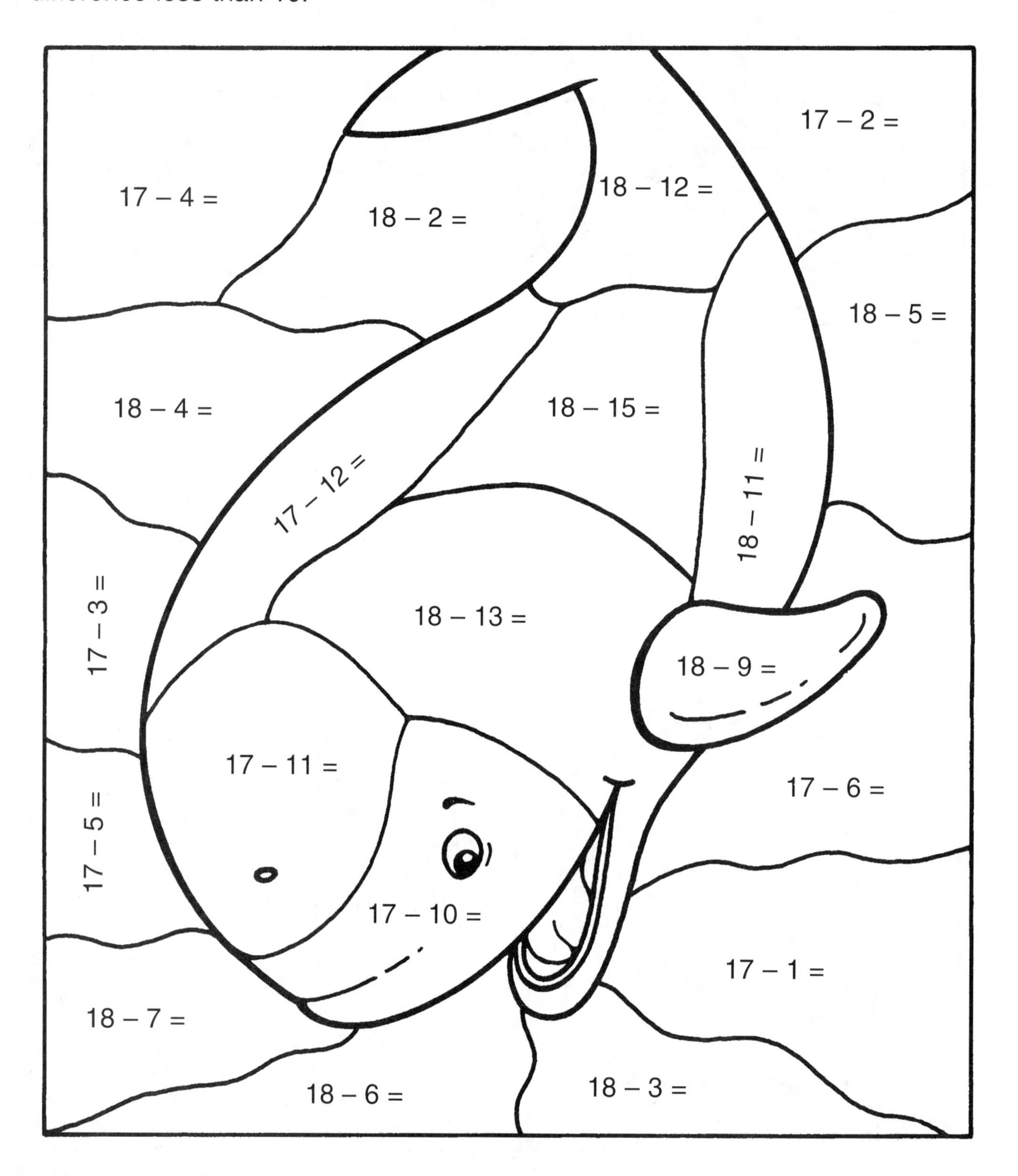

Practice 22

Begin with the number in the center. Subtract one of the numbers in the middle ring from the number in the center, and write the difference in the outer ring. The first one has been done for you. Repeat until you have filled every space in the outer rings.

1.

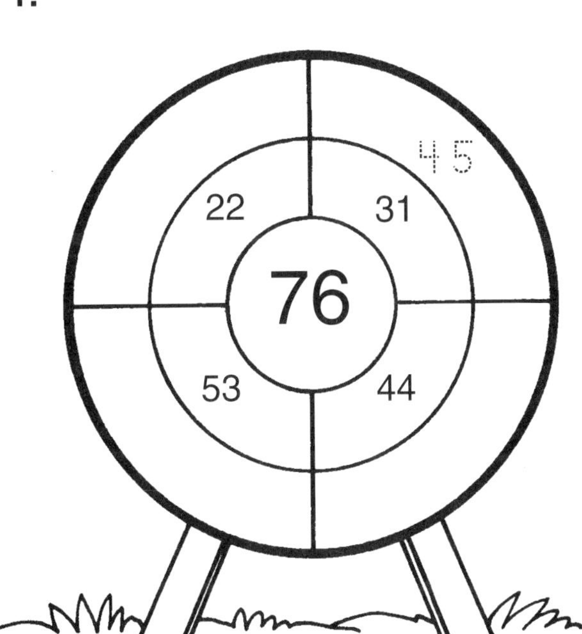

2.

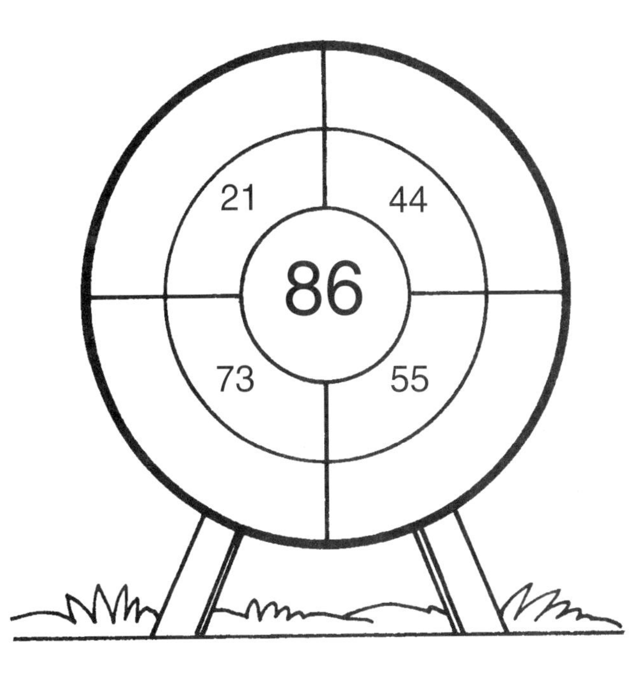

3.

4.

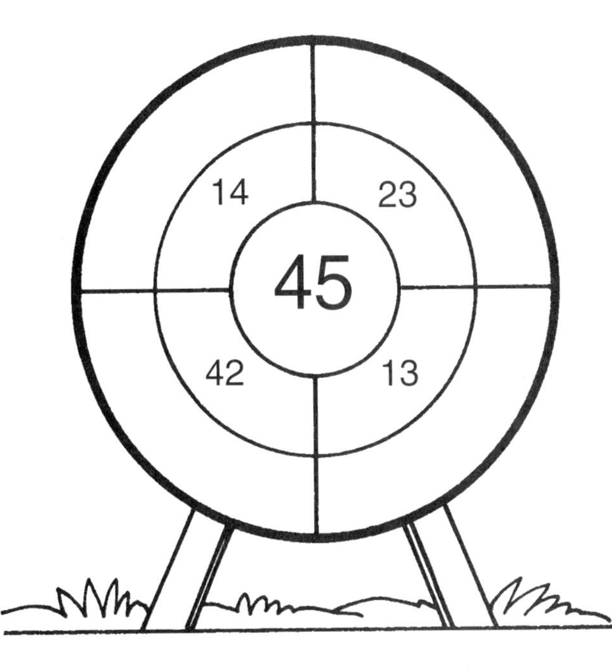

Practice 23

Find the answers. Color the pictures.

41 = yellow	46 = red	51 = blue
56 = green	61 = brown	

1.

2.

3.

4.

5.

6.

7.

8.

9.

Practice 24

Cross out each answer in the mitt as you solve the problems.

1.
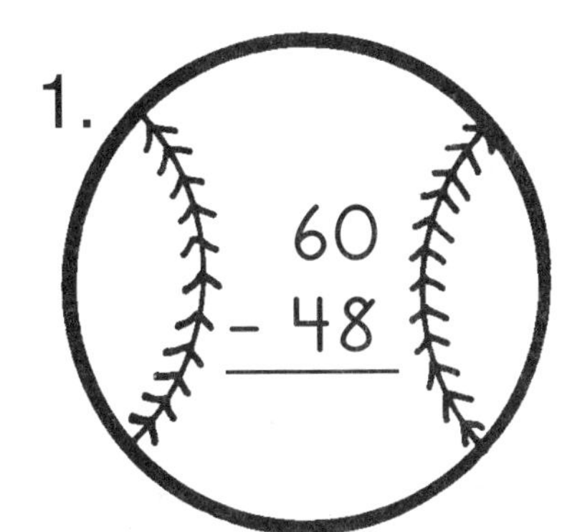

2.
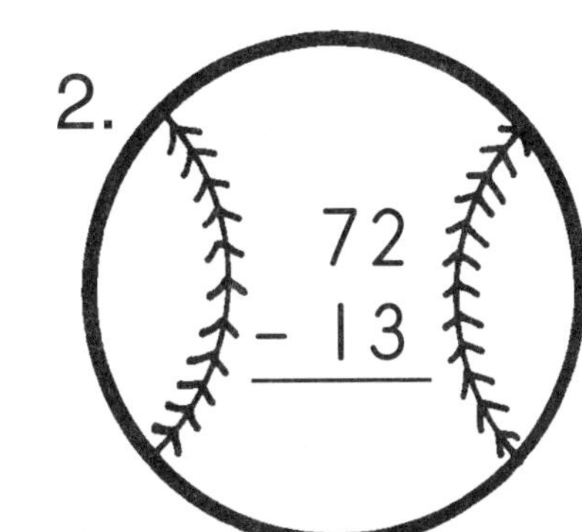

3.
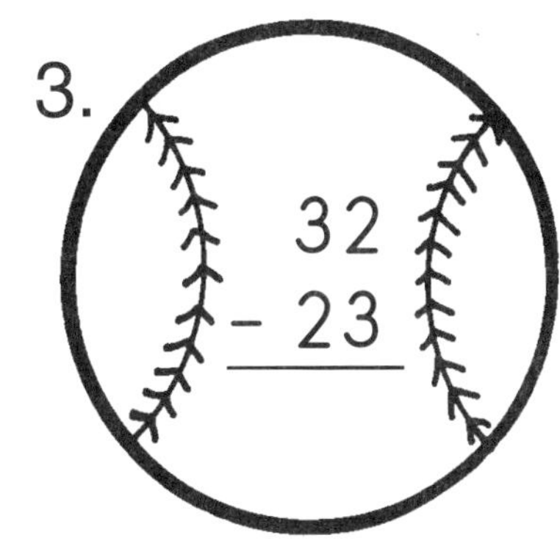

4.
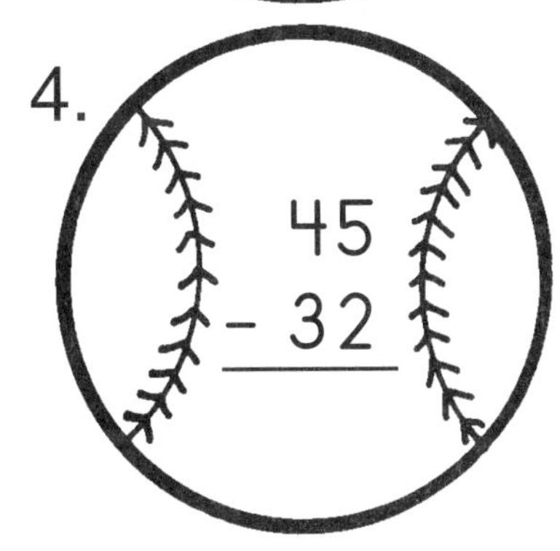

5.
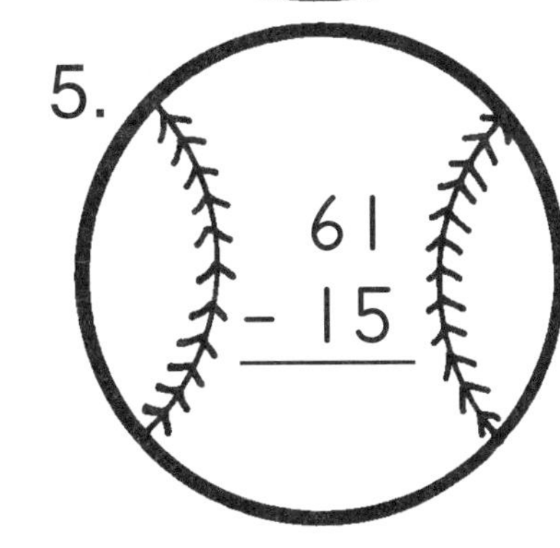

6.
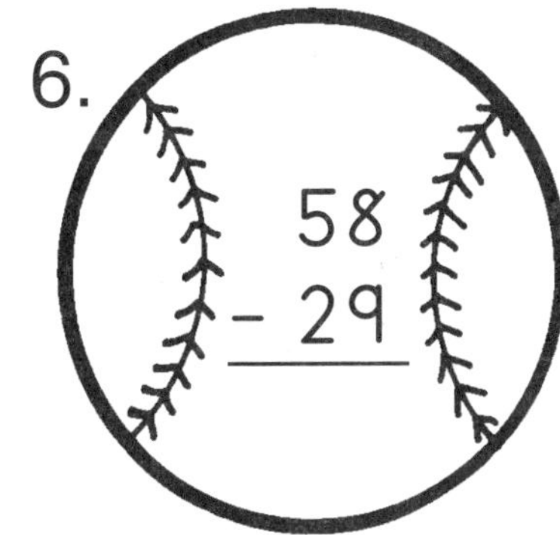

7.
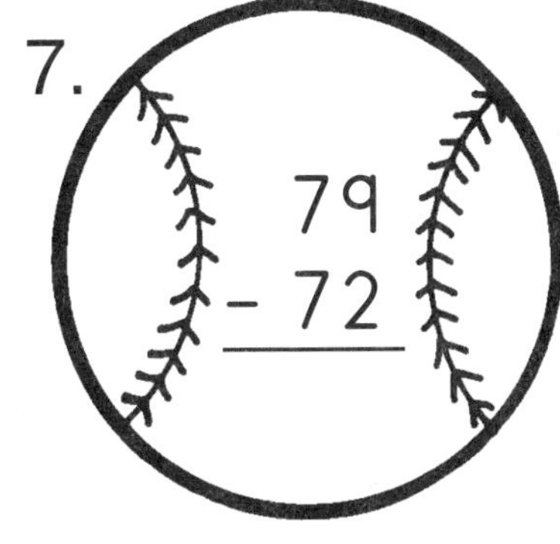

8.
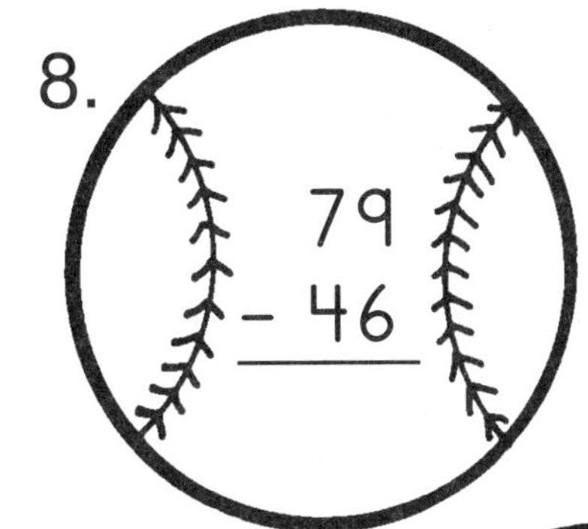

9.
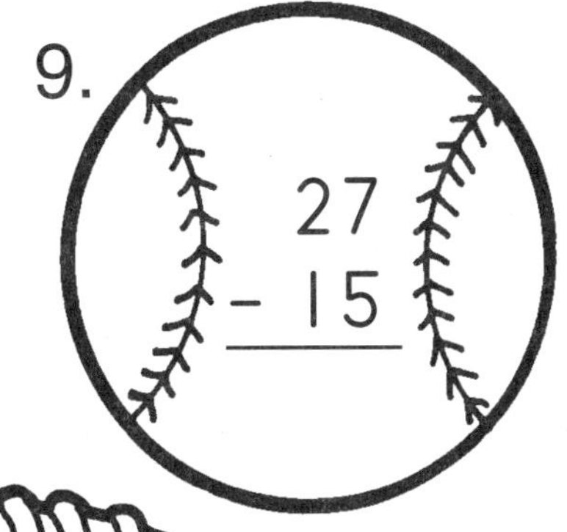

10.
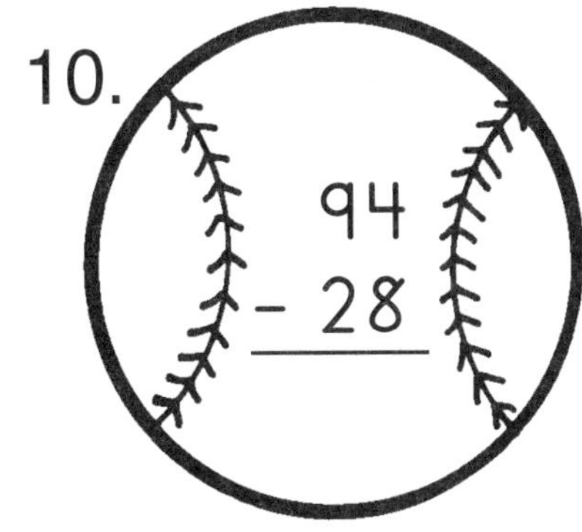

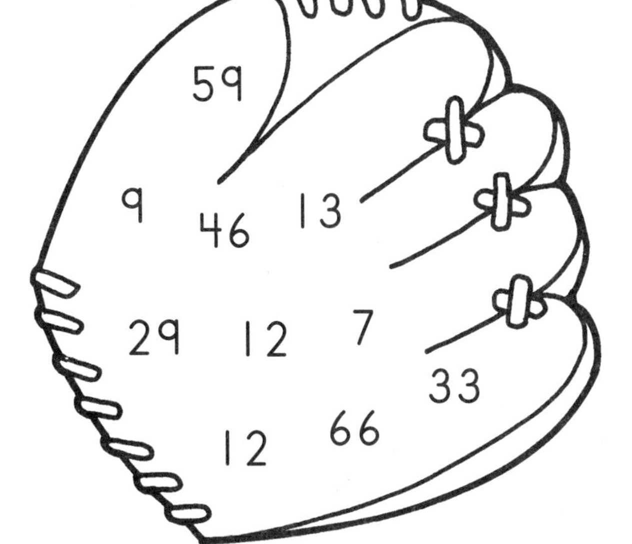

Practice 25

To find out which bear ate the cookies, solve each subtraction problem. The bear with the numeral 6 in the difference took the cookies.

Who ate the cookies?

Practice 26            

Guess what is in the box. Find the answers. Then write the letter in each box that matches each answer. Read the word you spell and draw it in the box.

20	21	22	23	24	25
g	e	a	t	i	r

44	56	50	62	75	74
−22	−33	−26	−42	−54	−49
22					
a	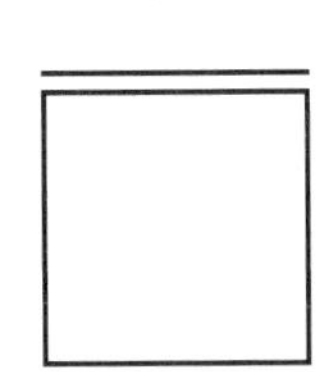		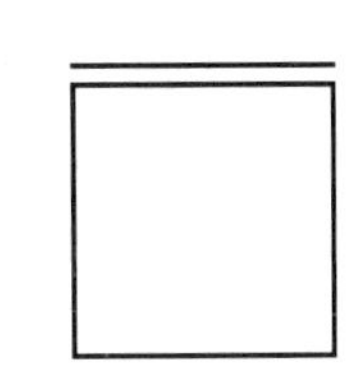	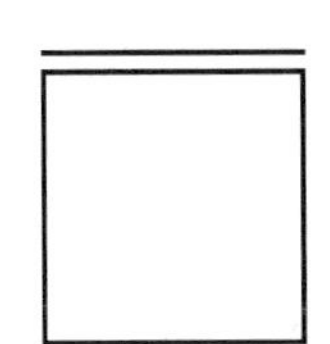	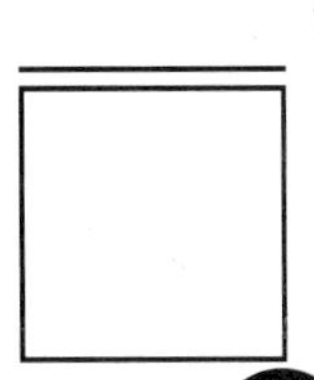

Practice 27

Find the answers. Color the pictures.

25 = orange   33 = green   35 = yellow
42 = purple   45 = blue

Practice 28

Guess what is in the box. Find the answers. Then write the letter in each box that matches each answer. Read the word you spell and draw it in the box.

35	36	37	38	39	40
a	c	j	e	k	t

75	75	92	58	72	99	87
-40	-38	-57	-22	-33	-61	-47
35						
a						

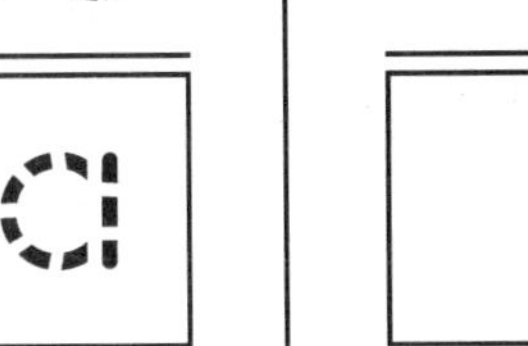  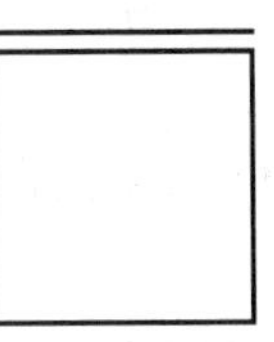 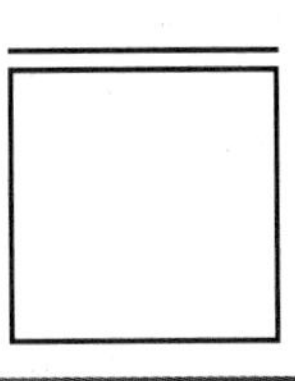 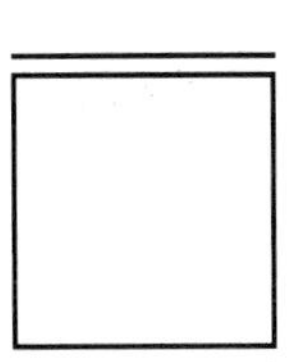 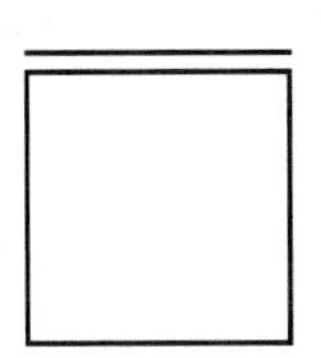 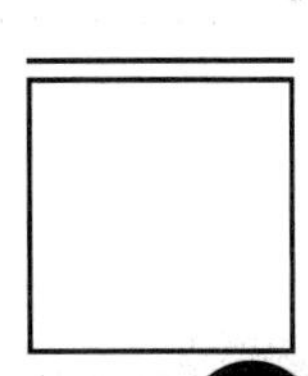

Practice 29           

Subtract the number on the second car from the number on the engine. Then subtract the number on the third car from the difference between those numbers. Continue subtracting until you get to the last car.

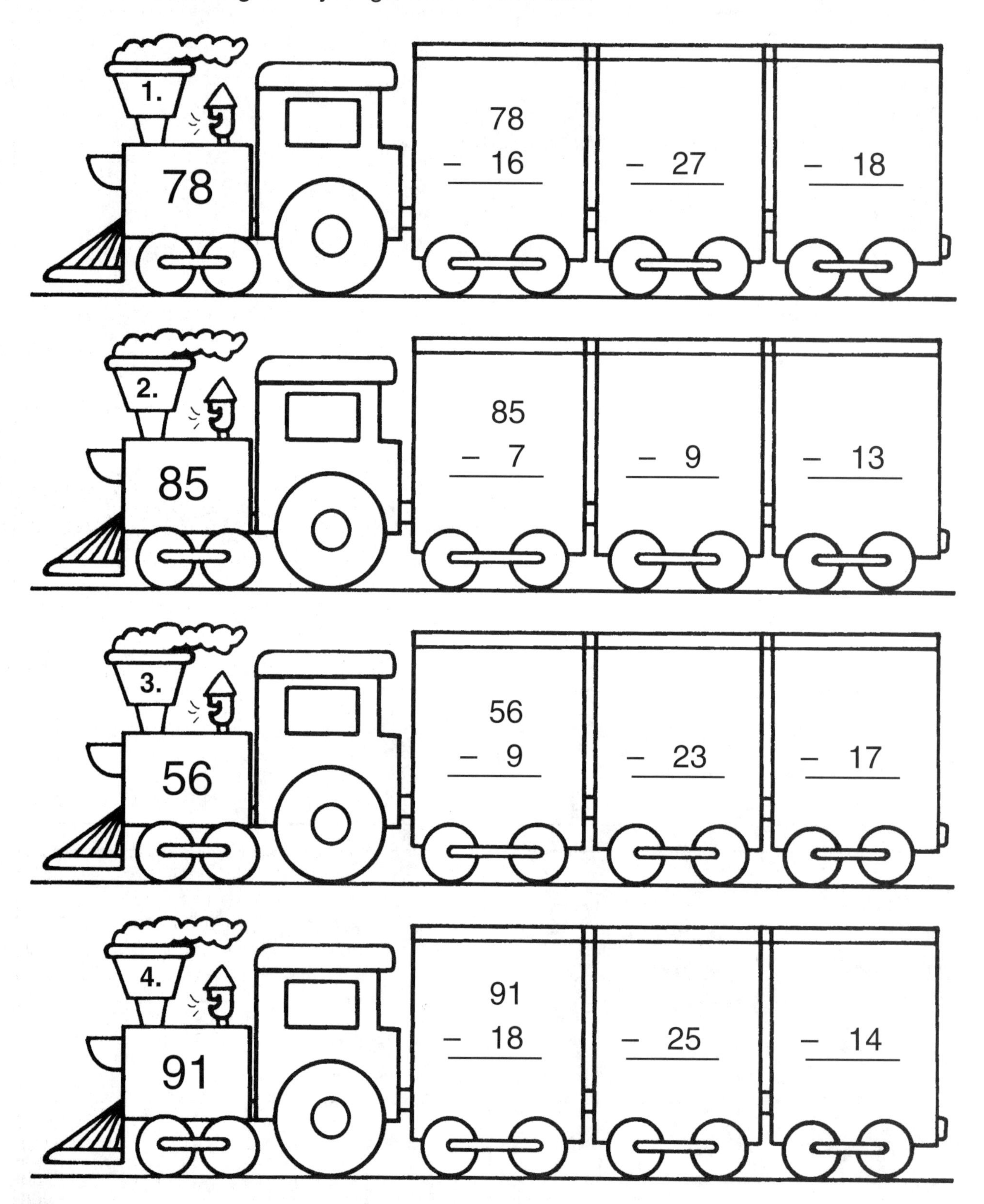

Practice 30

Solve each subtraction problem. To win this tic-tac-toe game, draw a line through the three squares that have a 7 in their answers.

$66 - 39$	$97 - 28$	$37 - 19$
$58 - 19$	$85 - 68$	$73 - 18$
$72 - 19$	$39 - 18$	$95 - 18$

Practice 31

Guess what is in the box. Find the answers. Then write the letter in each box that matches each answer. Read the word you spell and draw it in the box.

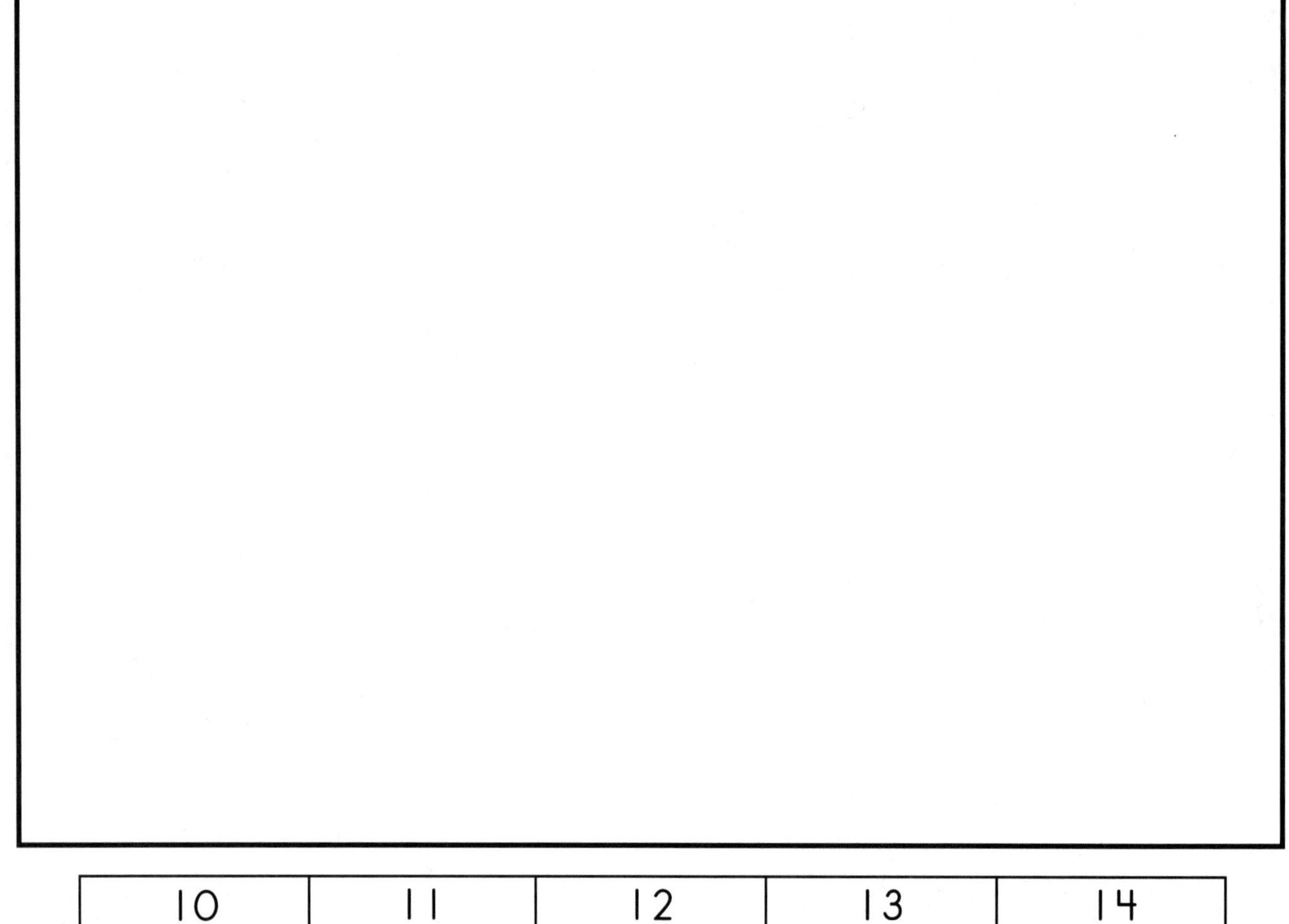

10	11	12	13	14
j	b	a	n	o

42	33	46	54	77	28
−30	−22	−34	−41	−67	−14
12					
	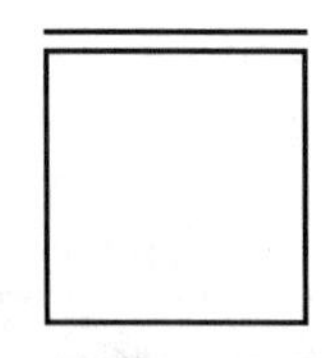	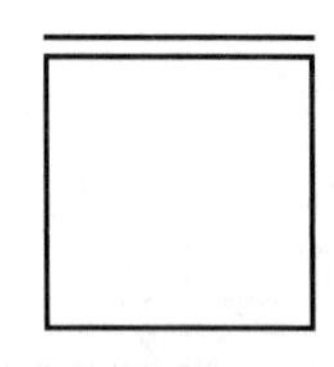	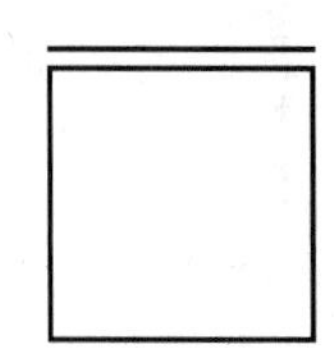	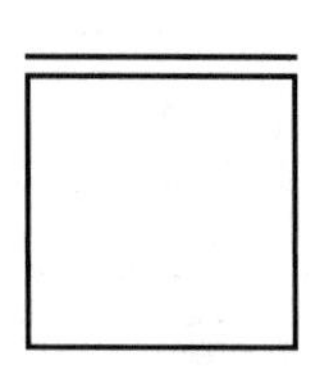	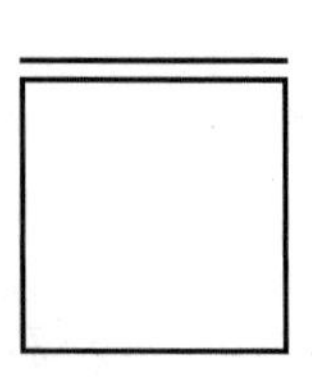

Practice 32

Read each word problem. Write the number sentence it shows. Find the difference.

1.

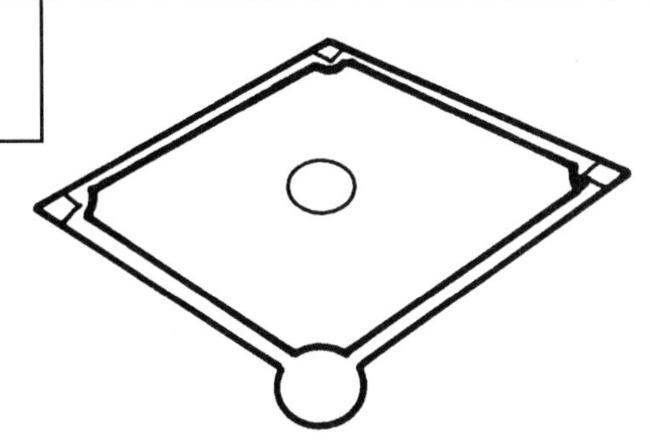

At the first baseball game of the season, 94 fans came to watch. During the second game, there were 76 fans. How many fewer fans came to watch the second game?

$94 - 76 = 18$

2.

Kim had 42 dolls in her collection. She gave away 26 dolls. How many did she have left?

3.

Mr. Jones is 62 years old. Mrs. Harris is 59. What is the difference in their ages?

4.

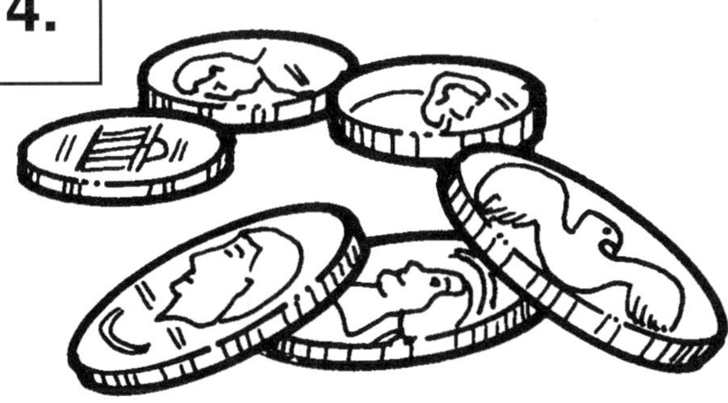

Casey has 95 coins in her collection. Alice has 42. What is the difference between the number of coins each has?

Practice 33

Read each word problem. Write the number sentence it shows. Find the difference.

1.

Farmer Cole raised 96 bushels of wheat. Farmer Dale raised 78 bushels. What is the difference in the number of bushels each raised?

96 − 78 = 18

2.

Dennis scored 43 points in his basketball game. Claire scored 20. What was the difference in points each earned?

3.

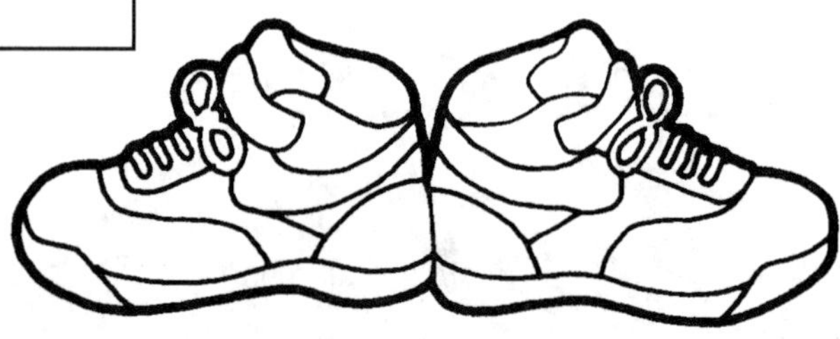

Jason bought a pair of shoes for 63 dollars. Clark bought a pair for 48 dollars. What was the difference paid?

4.

Jill counted 63 ants near an ant hill. Jack counted 40. What is the difference in the ants counted?

Practice 34           

Find the differences.

1. 49 - 28	**7.** 74 - 72	**13.** 74 - 68	**19.** 97 - 50
2. 97 - 35	**8.** 50 - 22	**14.** 38 - 15	**20.** 45 - 29
3. 30 - 16	**9.** 41 - 32	**15.** 45 - 29	**21.** 34 - 17
4. 56 - 26	**10.** 37 - 30	**16.** 79 - 32	**22.** 74 - 19
5. 40 - 33	**11.** 72 - 53	**17.** 60 - 14	**23.** 28 - 28
6. 86 - 56	**12.** 77 - 70	**18.** 86 - 16	**24.** 55 - 54

Practice 35

Find the differences.

1. 51 − 50	7. 69 − 12	13. 69 − 16	19. 36 − 13
2. 64 − 42	8. 72 − 38	14. 71 − 59	20. 89 − 80
3. 94 − 23	9. 48 − 18	15. 68 − 13	21. 51 − 17
4. 27 − 10	10. 52 − 11	16. 96 − 41	22. 91 − 19
5. 78 − 52	11. 19 − 15	17. 82 − 30	23. 46 − 31
6. 67 − 14	12. 62 − 31	18. 93 − 90	24. 87 − 43

Practice 36

Find the differences.

1. 31 − 23	**7.** 79 − 32	**13.** 85 − 21	**19.** 69 − 37
2. 75 − 42	**8.** 57 − 51	**14.** 51 − 20	**20.** 98 − 34
3. 54 − 23	**9.** 88 − 44	**15.** 42 − 28	**21.** 87 − 28
4. 42 − 26	**10.** 63 − 23	**16.** 71 − 56	**22.** 69 − 43
5. 88 − 26	**11.** 86 − 14	**17.** 36 − 32	**23.** 46 − 41
6. 61 − 33	**12.** 53 − 32	**18.** 97 − 60	**24.** 77 − 63

Test Practice 1

1.  – = ______

Ⓐ 5 Ⓑ 3 Ⓒ 2 Ⓓ 1

2. 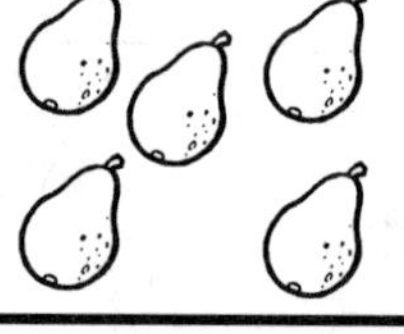 – = ______

Ⓐ 2 Ⓑ 1 Ⓒ 5 Ⓓ 3

3. 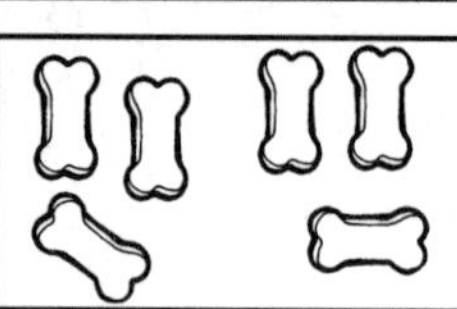 – 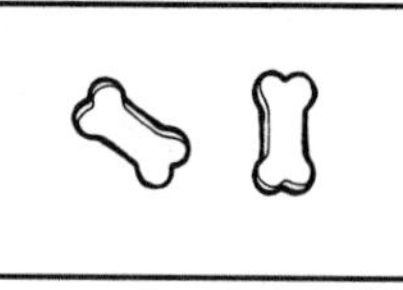 = ______

Ⓐ 0 Ⓑ 2 Ⓒ 4 Ⓓ 6

4. 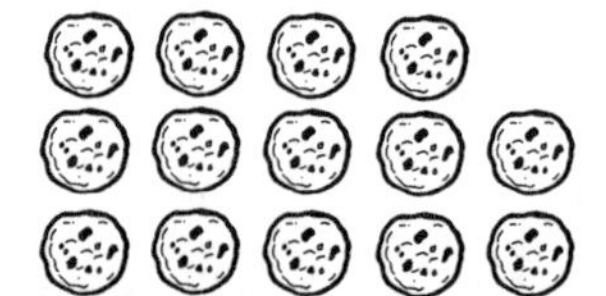 –  = ______

Ⓐ 17 Ⓑ 10 Ⓒ 13 Ⓓ 11

5.  –  = ______

Ⓐ 1 Ⓑ 2 Ⓒ 4 Ⓓ 6

6. 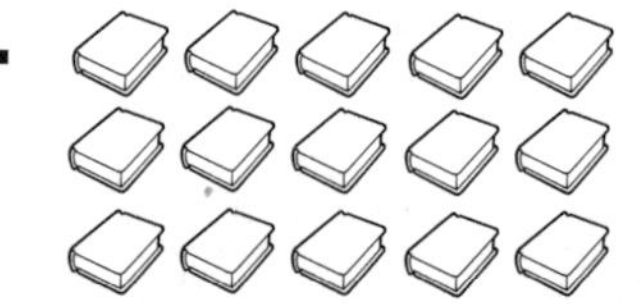 –  = ______

Ⓐ 12 Ⓑ 13 Ⓒ 16 Ⓓ 11

7. 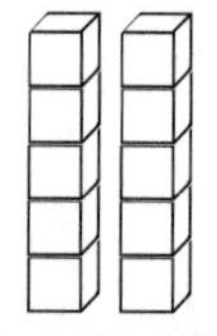 – 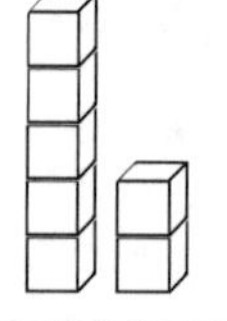 = ______

Ⓐ 3 Ⓑ 7 Ⓒ 10 Ⓓ 13

8. 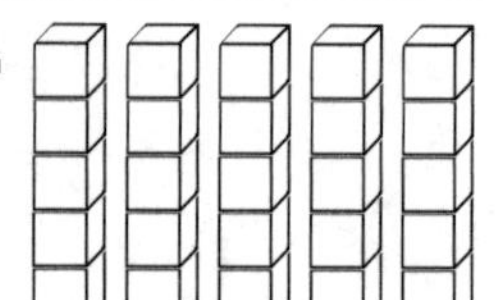 – 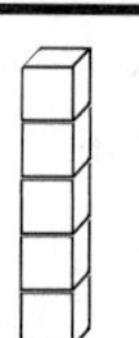 = ______

Ⓐ 20 Ⓑ 15 Ⓒ 30 Ⓓ 5

9.  – = ______

Ⓐ 9 Ⓑ 7 Ⓒ 5 Ⓓ 0

Test Practice 2

1.

$$\begin{array}{r} 5 \\ +\ 3 \\ \hline \end{array}$$

Ⓐ 53
Ⓑ 8
Ⓒ 9
Ⓓ 2

2.

$$\begin{array}{r} 41 \\ +\ 12 \\ \hline \end{array}$$

Ⓐ 18
Ⓑ 8
Ⓒ 15
Ⓓ 53

3.

3 + 3 + 2 =

Ⓐ 9
Ⓑ 18
Ⓒ 80
Ⓓ 8

4.

$$\begin{array}{r} 42 \\ +\ 22 \\ \hline \end{array}$$

Ⓐ 22
Ⓑ 60
Ⓒ 64
Ⓓ 20

5.

$$\begin{array}{r} 3 \\ 4 \\ +\ 2 \\ \hline \end{array}$$

Ⓐ 9
Ⓑ 12
Ⓒ 11
Ⓓ 7

6.

$$\begin{array}{r} 56 \\ +\ 3 \\ \hline \end{array}$$

Ⓐ 86
Ⓑ 88
Ⓒ 53
Ⓓ 59

7.

1 + 6 + 2 =

Ⓐ 7
Ⓑ 10
Ⓒ 12
Ⓓ 9

8.

6 + 3 + 0 =

Ⓐ 9
Ⓑ 8
Ⓒ 6
Ⓓ 63

9.

$$\begin{array}{r} 42 \\ +\ 3 \\ \hline \end{array}$$

Ⓐ 45
Ⓑ 52
Ⓒ 40
Ⓓ 67

10.

$$\begin{array}{r} 35 \\ +\ 2 \\ \hline \end{array}$$

Ⓐ 77
Ⓑ 17
Ⓒ 37
Ⓓ 73

11.

$$\begin{array}{r} 53 \\ +\ 45 \\ \hline \end{array}$$

Ⓐ 8
Ⓑ 98
Ⓒ 18
Ⓓ 89

12.

$$\begin{array}{r} 40 \\ +\ 10 \\ \hline \end{array}$$

Ⓐ 30
Ⓑ 40
Ⓒ 50
Ⓓ 10

Test Practice 3

1. $\begin{array}{r} 6 \\ +\ 5 \\ \hline \end{array}$

Ⓐ 13
Ⓑ 11
Ⓒ 1
Ⓓ 12

2. $\begin{array}{r} 7 \\ +\ 4 \\ \hline \end{array}$

Ⓐ 1
Ⓑ 10
Ⓒ 11
Ⓓ 9

3. $\begin{array}{r} 15 \\ +\ 26 \\ \hline \end{array}$

Ⓐ 31
Ⓑ 38
Ⓒ 41
Ⓓ 48

4. $7 + 6 =$

Ⓐ 1
Ⓑ 13
Ⓒ 42
Ⓓ 76

5. $\begin{array}{r} 87 \\ +\ 45 \\ \hline \end{array}$

Ⓐ 132
Ⓑ 122
Ⓒ 120
Ⓓ 142

6. $\begin{array}{r} 47 \\ 35 \\ +\ \ 5 \\ \hline \end{array}$

Ⓐ 77
Ⓑ 83
Ⓒ 87
Ⓓ 97

7. $42 + 28 =$

Ⓐ 16
Ⓑ 60
Ⓒ 70
Ⓓ 72

8. $\begin{array}{r} 123 \\ +\ 797 \\ \hline \end{array}$

Ⓐ 820
Ⓑ 810
Ⓒ 910
Ⓓ 920

9. $\begin{array}{r} 5 \\ 7 \\ +\ 3 \\ \hline \end{array}$

Ⓐ 10
Ⓑ 12
Ⓒ 15
Ⓓ 16

10. $\begin{array}{r} 2096 \\ +\ 5932 \\ \hline \end{array}$

Ⓐ 8968
Ⓑ 8978
Ⓒ 8028
Ⓓ 8868

11. $\begin{array}{r} 90 \\ +\ 60 \\ \hline \end{array}$

Ⓐ 30
Ⓑ 96
Ⓒ 130
Ⓓ 150

12. $\begin{array}{r} 9 \\ +\ 5 \\ \hline \end{array}$

Ⓐ 4
Ⓑ 14
Ⓒ 59
Ⓓ 95

Test Practice 4

1.

$$\begin{array}{r} 6 \\ -\ 4 \\ \hline \end{array}$$

Ⓐ 10
Ⓑ 6
Ⓒ 4
Ⓓ 2

2.

$$\begin{array}{r} 8 \\ -\ 2 \\ \hline \end{array}$$

Ⓐ 6
Ⓑ 8
Ⓒ 10
Ⓓ 2

3.

$$\begin{array}{r} 12 \\ -\ 8 \\ \hline \end{array}$$

Ⓐ 20
Ⓑ 16
Ⓒ 4
Ⓓ 8

4.

$$\begin{array}{r} 45 \\ -\ \ 4 \\ \hline \end{array}$$

Ⓐ 49
Ⓑ 10
Ⓒ 41
Ⓓ 44

5.

$$\begin{array}{r} 55 \\ -\ 11 \\ \hline \end{array}$$

Ⓐ 11
Ⓑ 66
Ⓒ 44
Ⓓ 55

6.

$12 - 6 =$

Ⓐ 5
Ⓑ 6
Ⓒ 10
Ⓓ 18

7.

$$\begin{array}{r} 77 \\ -\ \ 6 \\ \hline \end{array}$$

Ⓐ 17
Ⓑ 11
Ⓒ 71
Ⓓ 61

8.

$$\begin{array}{r} 58 \\ -\ 15 \\ \hline \end{array}$$

Ⓐ 33
Ⓑ 43
Ⓒ 63
Ⓓ 73

9.

$$\begin{array}{r} 60 \\ -\ 10 \\ \hline \end{array}$$

Ⓐ 70
Ⓑ 60
Ⓒ 50
Ⓓ 10

10.

$$\begin{array}{r} 56 \\ -\ \ 3 \\ \hline \end{array}$$

Ⓐ 49
Ⓑ 43
Ⓒ 53
Ⓓ 59

11.

$9 - 9 =$

Ⓐ 18
Ⓑ 10
Ⓒ 9
Ⓓ 0

12.

$$\begin{array}{r} 38 \\ -\ \ 5 \\ \hline \end{array}$$

Ⓐ 43
Ⓑ 33
Ⓒ 23
Ⓓ 15

Test Practice 5

1.
$$\begin{array}{r} 82 \\ -\ 56 \\ \hline \end{array}$$
Ⓐ 136
Ⓑ 36
Ⓒ 26
Ⓓ 32

2.
$$\begin{array}{r} 8 \\ -\ 6 \\ \hline \end{array}$$
Ⓐ 14
Ⓑ 12
Ⓒ 4
Ⓓ 2

3.
14 – 8 =
Ⓐ 22
Ⓑ 14
Ⓒ 8
Ⓓ 6

4.
$$\begin{array}{r} 52 \\ -\ 47 \\ \hline \end{array}$$
Ⓐ 5
Ⓑ 15
Ⓒ 89
Ⓓ 99

5.
$$\begin{array}{r} 75 \\ -\ \ 6 \\ \hline \end{array}$$
Ⓐ 79
Ⓑ 69
Ⓒ 71
Ⓓ 61

6.
$$\begin{array}{r} 809 \\ -\ 215 \\ \hline \end{array}$$
Ⓐ 514
Ⓑ 504
Ⓒ 594
Ⓓ 694

7.
63 – 8 =
Ⓐ 71
Ⓑ 75
Ⓒ 55
Ⓓ 65

8.
$$\begin{array}{r} 252 \\ -\ 136 \\ \hline \end{array}$$
Ⓐ 116
Ⓑ 126
Ⓒ 388
Ⓓ 398

9.
$$\begin{array}{r} 764 \\ -\ 569 \\ \hline \end{array}$$
Ⓐ 195
Ⓑ 115
Ⓒ 205
Ⓓ 215

10.
$$\begin{array}{r} 719 \\ -\ 428 \\ \hline \end{array}$$
Ⓐ 311
Ⓑ 301
Ⓒ 291
Ⓓ 201

11.
50 – 5 =
Ⓐ 250
Ⓑ 55
Ⓒ 45
Ⓓ 35

12.
$$\begin{array}{r} 26 \\ -\ \ 7 \\ \hline \end{array}$$
Ⓐ 19
Ⓑ 21
Ⓒ 23
Ⓓ 29

Test Practice 6

1.

$$\begin{array}{r} 21 \\ +\ 11 \\ \hline \end{array}$$

Ⓐ 30
Ⓑ 31
Ⓒ 32
Ⓓ 33

2.

$$\begin{array}{r} 43 \\ +\ 12 \\ \hline \end{array}$$

Ⓐ 45
Ⓑ 50
Ⓒ 55
Ⓓ 60

3.

$$\begin{array}{r} 120 \\ +\ 134 \\ \hline \end{array}$$

Ⓐ 250
Ⓑ 252
Ⓒ 253
Ⓓ 254

4.

$$\begin{array}{r} 346 \\ +\ 178 \\ \hline \end{array}$$

Ⓐ 522
Ⓑ 532
Ⓒ 524
Ⓓ 534

5.

$$\begin{array}{r} 167 \\ +\ 240 \\ \hline \end{array}$$

Ⓐ 417
Ⓑ 407
Ⓒ 517
Ⓓ 507

6.

$$\begin{array}{r} 780 \\ +\ 125 \\ \hline \end{array}$$

Ⓐ 900
Ⓑ 905
Ⓒ 910
Ⓓ 915

7.

$$\begin{array}{r} 34 \\ -\ 11 \\ \hline \end{array}$$

Ⓐ 23
Ⓑ 33
Ⓒ 43
Ⓓ 13

8.

$$\begin{array}{r} 67 \\ -\ 20 \\ \hline \end{array}$$

Ⓐ 37
Ⓑ 27
Ⓒ 47
Ⓓ 57

9.

$$\begin{array}{r} 46 \\ -\ 39 \\ \hline \end{array}$$

Ⓐ 7
Ⓑ 8
Ⓒ 9
Ⓓ 6

10.

$$\begin{array}{r} 135 \\ -\ 119 \\ \hline \end{array}$$

Ⓐ 16
Ⓑ 18
Ⓒ 17
Ⓓ 19

11.

$$\begin{array}{r} 562 \\ -\ 499 \\ \hline \end{array}$$

Ⓐ 62
Ⓑ 63
Ⓒ 64
Ⓓ 67

12.

$$\begin{array}{r} 341 \\ -\ 112 \\ \hline \end{array}$$

Ⓐ 219
Ⓑ 239
Ⓒ 229
Ⓓ 249

Answer Sheet         

Test Practice 1	Test Practice 2	Test Practice 3
1. (A) (B) (C) (D)	1. (A) (B) (C) (D)	1. (A) (B) (C) (D)
2. (A) (B) (C) (D)	2. (A) (B) (C) (D)	2. (A) (B) (C) (D)
3. (A) (B) (C) (D)	3. (A) (B) (C) (D)	3. (A) (B) (C) (D)
4. (A) (B) (C) (D)	4. (A) (B) (C) (D)	4. (A) (B) (C) (D)
5. (A) (B) (C) (D)	5. (A) (B) (C) (D)	5. (A) (B) (C) (D)
6. (A) (B) (C) (D)	6. (A) (B) (C) (D)	6. (A) (B) (C) (D)
7. (A) (B) (C) (D)	7. (A) (B) (C) (D)	7. (A) (B) (C) (D)
8. (A) (B) (C) (D)	8. (A) (B) (C) (D)	8. (A) (B) (C) (D)
9. (A) (B) (C) (D)	9. (A) (B) (C) (D)	9. (A) (B) (C) (D)
	10. (A) (B) (C) (D)	10. (A) (B) (C) (D)
	11. (A) (B) (C) (D)	11. (A) (B) (C) (D)
	12. (A) (B) (C) (D)	12. (A) (B) (C) (D)

Test Practice 4	Test Practice 5	Test Practice 6
1. (A) (B) (C) (D)	1. (A) (B) (C) (D)	1. (A) (B) (C) (D)
2. (A) (B) (C) (D)	2. (A) (B) (C) (D)	2. (A) (B) (C) (D)
3. (A) (B) (C) (D)	3. (A) (B) (C) (D)	3. (A) (B) (C) (D)
4. (A) (B) (C) (D)	4. (A) (B) (C) (D)	4. (A) (B) (C) (D)
5. (A) (B) (C) (D)	5. (A) (B) (C) (D)	5. (A) (B) (C) (D)
6. (A) (B) (C) (D)	6. (A) (B) (C) (D)	6. (A) (B) (C) (D)
7. (A) (B) (C) (D)	7. (A) (B) (C) (D)	7. (A) (B) (C) (D)
8. (A) (B) (C) (D)	8. (A) (B) (C) (D)	8. (A) (B) (C) (D)
9. (A) (B) (C) (D)	9. (A) (B) (C) (D)	9. (A) (B) (C) (D)
10. (A) (B) (C) (D)	10. (A) (B) (C) (D)	10. (A) (B) (C) (D)
11. (A) (B) (C) (D)	11. (A) (B) (C) (D)	11. (A) (B) (C) (D)
12. (A) (B) (C) (D)	12. (A) (B) (C) (D)	12. (A) (B) (C) (D)

Answer Key

Page 4
1. 12
2. 10
3. 11
4. 8
5. 6
6. 7
7. 9
8. 8
9. 10
10. 12
11. 14
12. 9
13. 11
14. 13
15. 12
16. 17
17. 16
18. 20
19. 15
20. 24

Page 5
1. 20, 21, 22, 23, 24, 25, 26, 27, 28, 29, 30
2. 30, 31, 32, 33, 34, 35, 36, 37, 38, 39, 40
3. 40, 41, 42, 43, 44, 45, 46, 47, 48, 49, 50
4. 50, 51, 52, 53, 54, 55, 56, 57, 58, 59, 60
5. 60, 61, 62, 63, 64, 65, 66, 67, 68, 69, 70
6. 70, 71, 72, 73, 74, 75, 76, 77, 78, 79, 80
7. 80, 81, 82, 83, 84, 85, 86, 87, 88, 89, 90
8. 90, 91, 92, 93, 94, 95, 96, 97, 98, 99, 100

Page 6
1. 41
2. 38
3. 24
4. 57
5. 24
6. 57
7. 24
8. 38
9. 41

Page 7
1. 33
2. 29
3. 43
4. 29
5. 34
6. 53
7. 64
8. 62
9. 59

Page 8
1. 39
2. 47
3. 52
4. 72
5. 24
6. 27

Secret number: 27

Page 9
27; a
26; c
29; l
28; o
26; c
25; k
a clock

Page 10
a. 42
b. 35
c. 61
d. 29
e. 20
g. 37
i. 51
l. 53
m. 92
n. 84
o. 67
r. 31
s. 66
t. 25
u. 39
w. 48
y. 72

Congratulations! You are adding two-digit numbers.

Page 11
50; a
52; h
51; o
53; r
48; s
49; e
a horse

Page 12
1. 71
2. 43
3. 86
4. 60
5. 67
6. 81
7. 62

Page 13
82; a
84; c
85; r
83; o
84; c
83; o
80; d
87; i
86; l
81; e
a crocodile

Page 14
1. 85 + 77 = 162 cents
2. 54 + 99 = 153 cents
3. 72 + 77 = 149 cents
4. 54 + 85 = 139 cents
5. 99 + 77 + 54 = 230 cents
6. 72 + 54 + 85 = 211 cents

Page 15
1. 49 + 57 = 106 dollars
2. 26 + 32 = 58 dollars
3. 17 + 64 = 81 dollars
4. 32 + 57 = 89 dollars
5. 64 + 17 + 49 = 130 dollars
6. 26 + 57 + 32 = 115 dollars

Page 16
1. 16 + 14 + 12 = 42
2. 18 + 22 + 36 = 76
3. 17 + 13 + 42 = 72
4. 15 + 23 + 24 = 62

Page 17
1. 53 + 24 + 12 = 89
2. 57 + 24 + 43 = 124
3. 24 + 32 + 34 = 90
4. 36 + 40 + 67 = 143

Page 18
1. 77
2. 132
3. 46
4. 82
5. 73
6. 142
7. 89
8. 72
9. 93
10. 57
11. 125
12. 137
13. 142
14. 53
15. 74
16. 66
17. 74
18. 102
19. 147
20. 74
21. 51
22. 93
23. 55
24. 109

Page 19
1. 61
2. 106
3. 117
4. 37
5. 130
6. 81
7. 81
8. 110
9. 66
10. 63
11. 34
12. 93
13. 85
14. 130
15. 81
16. 137
17. 112
18. 183
19. 49
20. 109
21. 68
22. 110
23. 77
24. 130

Page 20
1. 54
2. 117
3. 77
4. 68
5. 150
6. 94
7. 111
8. 108
9. 132
10. 87
11. 100
12. 85
13. 106
14. 71
15. 60
16. 127
17. 68
18. 157
19. 106
20. 132
21. 115
22. 112
23. 87
24. 140

Page 21
1. 143
2. 131
3. 91
4. 117
5. 186
6. 157
7. 123
8. 163
9. 145
10. 129
11. 185
12. 106
13. 156
14. 100
15. 93
16. 132
17. 201
18. 174
19. 150
20. 182
21. 174
22. 119
23. 111
24. 170

Page 22
1. 10, 9, 8, 7, 6, 5, 4, 3, 2, 1, 0
2. 20, 19, 18, 17, 16, 15, 14, 13, 12, 11, 10
3. 30, 29, 28, 27, 26, 25, 24, 23, 22, 21, 20
4. 40, 39, 38, 37, 36, 35, 34, 33, 32, 31, 30
5. 50, 49, 48, 47, 46, 45, 44, 43, 42, 41, 40
6. 60, 59, 58, 57, 56, 55, 54, 53, 52, 51, 50
7. 70, 69, 68, 67, 66, 65, 64, 63, 62, 61, 60
8. 80, 79, 78, 77, 76, 75, 74, 73, 72, 71, 70

Answer Key

Page 23
13 – 10 = 3; yellow
13 – 9 = 4; yellow
13 – 6 = 7; red
13 – 5 = 8; red
13 – 3 = 10; green
13 – 7 = 6; yellow
13 – 9 = 4; yellow
13 – 4 = 9; green
13 – 12 = 1; blue
13 – 8 = 5; yellow
13 – 11 = 2; blue
13 – 1 = 12; white
13 – 2 = 11; white

Page 24
17 – 4 = 13; blue
18 – 2 = 16; blue
18 – 12 = 6; gray
17 – 2 = 15; blue
18 – 4 = 14; blue
17 – 12 = 5; gray
18 – 15 = 3; gray
18 – 11 = 7; gray
18 – 5 = 13; blue
17 – 3 = 14; blue
18 – 13 = 5; gray
18 – 9 = 9; gray
17 – 5 = 12; blue
17 – 11 = 6; gray
17 – 10 = 7; gray
17 – 6 = 11; blue
17 – 1 = 16; blue
18 – 7 = 11; blue
18 – 6 = 12; blue
18 – 3 = 15; blue

Page 25
Answers are listed left to right
1. 54, 45, 23, 32
2. 65, 42, 13, 31
3. 27, 11, 35, 46
4. 31, 22, 3, 32

Page 26
1. 46; red
2. 56; green
3. 41; yellow
4. 61; brown
5. 56; green
6. 51; blue
7. 61; brown
8. 46; red
9. 46; red

Page 27
1. 12
2. 59
3. 9
4. 13
5. 46
6. 29
7. 7
8. 33
9. 12
10. 66

Page 28
1. 13
2. 4
3. 18
4. 57
5. 16
6. 37

Bear #5 ate the cookies.

Page 29
22; a
23; t
24; i
20; g
21; e
25; r
a tiger

Page 30
73 – 48 = 25
84 – 49 = 35
90 – 48 = 42
61 – 16 = 45
62 – 29 = 33
80 – 38 = 42
93 – 48 = 45
82 – 47 = 35
70 – 45 = 25

Page 31
35; a
37; j
35; a
36; c
39; k
38; e
40; t
a jacket

Page 32
1. 78 – 16 = 62
   62 – 27 = 35
   35 – 18 = 17
2. 85 – 7 = 78
   78 – 9 = 69
   69 – 13 = 56
3. 56 – 9 = 47
   47 – 23 = 24
   24 – 17 = 7
4. 91 – 18 = 73
   73 – 25 = 48
   48 – 14 = 34

Page 33
66 – 39 = 27
97 – 28 = 69
37 – 19 = 18
58 – 19 = 39
85 – 68 = 17
73 – 18 = 55
72 – 19 = 53
39 – 18 = 21
95 – 18 = 77

Page 34
12; a
11; b
12; a
13; n
10; j
14; o
a banjo

Page 35
1. 94 – 76 = 18
2. 42 – 26 = 16
3. 62 – 59 = 3
4. 95 – 42 = 53

Page 36
1. 96 – 78 = 18
2. 33 – 20 = 13
3. 63 – 48 = 15
4. 63 – 40 = 23

Page 37
1. 21
2. 62
3. 14
4. 30
5. 7
6. 30
7. 2
8. 28
9. 9
10. 7
11. 19
12. 7
13. 6
14. 23
15. 16
16. 47
17. 46
18. 70
19. 47
20. 16
21. 17
22. 55
23. 0
24. 1

Page 38
1. 1
2. 22
3. 71
4. 17
5. 26
6. 53
7. 57
8. 34
9. 30
10. 41
11. 4
12. 31
13. 53
14. 12
15. 55
16. 55
17. 52
18. 3
19. 23
20. 9
21. 34
22. 72
23. 15
24. 44

Page 39
1. 8
2. 33
3. 31
4. 16
5. 62
6. 28
7. 47
8. 6
9. 44
10. 40
11. 72
12. 21
13. 64
14. 31
15. 14
16. 15
17. 4
18. 37
19. 32
20. 64
21. 59
22. 26
23. 5
24. 14

Page 40
1. B
2. A
3. C
4. B
5. C
6. A
7. A
8. A
9. D

Page 41
1. B
2. D
3. D
4. D
5. C
6. A
7. D
8. A
9. A
10. C
11. B
12. C

Page 42
1. B
2. C
3. C
4. C
5. B
6. A
7. C
8. D
9. C
10. C
11. D
12. B

Page 43
1. D
2. C
3. A
4. C
5. C
6. C
7. B
8. B
9. C
10. C
11. D
12. B

Page 44
1. C
2. C
3. D
4. D
5. A
6. B
7. C
8. A
9. A
10. C
11. C
12. A

Page 45
1. B
2. C
3. D
4. C
5. B
6. B
7. A
8. C
9. A
10. A
11. B
12. C